Welcome to the Eden Project – the living theatre of plants, people and possibilities

Thank you for visiting us and for buying this guide, which we hope will enrich your stay and act as a memento of your day with us. All profits from its sale go to the Eden Trust to further its educational, environmental and scientific aims.

The Eden Trust?

The Eden Trust is the registered charity that owns the Project – Destination Eden is the stage, if you like, on which we explore our interdependence with plants and ask what the future might look like. It's a matter of making choices, from the small steps we can all take, to global initiatives. Only one thing is certain. If that future isn't green we don't have one – so, together, we simply have to find ways to tread more lightly on this planet we share. We at Eden are taking our first faltering steps along this road, working with a fantastic group of partners and collaborators. Join us.

More than just a green theme park, then?

If we had created nothing more than a theme park – a diversionary entertainment away from the real world – the last few years would have been wasted. Nevertheless, we want this place to be an assault on the senses, vibrant, engaging and fun, or people aren't going to come; they're certainly not going to come back or tell their friends about us. Yes, Eden is about serious issues; and yes, it's meant to be educational in the widest sense. But most of all it is intended to be optimistic, celebrating what is good and positive about possible futures.

Possible futures? Such as?

The Times, in a leader article, once memorably described what we do here as 'education by visionaries'. More accurately, Eden is our vision of education – providing some answers, certainly, but mostly inspiring us to examine the world afresh. Expect conversations, stories, informed views, news and ideas, inviting you to make up your own mind.

So, we set out to make Eden fun and thoughtful, light-hearted yet capable of making a point – a place with serious ambitions and, in its creation, a symbol of the power of people working with the grain of nature.

First, it had to be a place of transformation – taking a site as sterile and derelict as only humans could conspire at and making it beautiful and fertile. We mixed up 85,000 tonnes of soil, largely from organic waste, and planted more than 1,000,000 plants to create the vast global garden you see today. No one has ever done this before.

Second, we wanted jaw-dropping, award-winning architecture from the best in the business, inspired by nature and crafted by human ingenuity and commitment – somewhere to remind us every day of the talent and potential of our fellow human beings.

Third, as we've developed this hole in the ground and grown our team to many hundred strong we've developed our values and the way we do business – the way we look after our staff, our suppliers, our partners and our neighbours. Sometimes it almost feels as if we've built a model world and we're on a journey of discovery to find out what we can do with it.

But Eden isn't a world in isolation; we are proud to be deeply rooted in the local community, without whose early and continuing support we simply wouldn't be here today. We bend over backwards to source locally most of what we need to run the place; currently over 50% of all our purchasing and 90% of our catering supplies come from Cornwall.

We are working with local companies to develop appropriate products to sell not just at Eden but also in high streets up and down the country – and perhaps beyond, sharing our luck, building what Eden stands for, spreading our message.

So, we're trying to make a difference, and, by coming today, you're helping us to do just that. We hope never to be so impertinent as to suggest that we can make massive changes on our own. Instead we simply try to promote a tone of voice – an attitude – that brings people together in an apolitical way and maybe encourages us all to think that idealistic need not be the same as naïve.

But we are still young and evolving. We hope you understand, and can feel part of that journey; please celebrate with us any successes you may find on site and don't cringe too much if we expose our feet of clay. We believe you only make mistakes if you're really trying!

0·01 Taking it from the top

When we bought Bodelva in 1998 it had just reached the end of its life as a china clay pit. It looked like a huge inverted cone, over 60 metres deep and the size of 35 football pitches – with no level ground, no soil, one or two gorse bushes and enough water to begin the Atlantis Project instead. Great start for a global garden! So we sliced 17 metres off the top and put it into the bottom to make the site better suited to people than mountain goats. It took twelve dumper trucks and eight bulldozers six months to shift 1.8 million tonnes of dirt.

The water: in the first two months it rained solidly and 43 million gallons of water drained into the pit. The engineers came up with a drainage system that could take anything the weather chucked at it. This subterranean masterpiece now collects all the water coming on to the site, on average 22 litres/sec or 20,000 bathfuls a day. The water is used to satisfy the needs of the plants, as irrigation, and the visitor, by flushing the loos. Rainwater that falls on the Biomes is used inside to create the misty atmosphere of the rainforest.

The earth: back on dry land, dodgy slopes were shaved back to a safe angle and terraces chopped out. Two thousand rock anchors, some up to 11 m long, were driven into the pit sides to stabilize them, and a 'soup' of plant seed and plant food sprayed on the slopes

to knit the surface together. We then poured in 85,000 tonnes of soil that we made from recycled waste and added over 5,000 types of plants.

The plants: most of our plants are not rare, except for the few that tell stories of rarity and the need for conservation. They were brought here to show us the raw materials of our lives: the plants we use every day. The plants weren't taken from the wild, either. Many were grown from seed in our nursery, which is currently growing plants for our next chapter (the Dry Tropics Biome). Others came from botanic gardens, research stations and supporters worldwide.

The story: the Eden Team merge fabulous horticulture with art, science and education to tell stories about us and our world. We need to find a balance between growing plants to meet our needs (and wants), and conserving the land worldwide. The plants in the Outdoor Biome come from all over the temperate world, including parts of Asia, America and the upper slopes of tropical mountains as well as Europe. Cornwall's mild climate helps. Yes, the green shrubs down to the left, near the giant wooden leaf, really are tea bushes. So come on in and explore, outdoors and in.

As well as the plants we hope you will like our collection of celebratory silk flags, inspired by various plant forms at Eden and designed, made and planted by Angus Watt. The entrance to the covered Biomes is via the link bridge between them. More about the Biome structures on page 54 and their stories on pages 19 and 38.

0.02 The flowerless garden: ancient plants

Ferns, horsetails and mosses were around before the dinosaurs roamed the earth. They thrive here on this north-facing slope. Look out for the copper ferns along the zigzag path sculpted and planted by Kate Munro and the Eden Green Team.

0.03 The plants that feed the world

There are over 250,000 plant species on the earth. But six, and a few close runners-up, have become the global crops that feed the world. Wheat, maize, rice and potatoes, grown outdoors here, are joined by bananas in the Humid Tropics Biome and pulses in the Warm Temperate. What do they have in common? They all produce starch, helping to provide energy and nutrition, and they can all be stored.

Maize is currently number one, in yield terms, eaten by people and their animals worldwide. Scientists have bred Quality Protein Maize with a better nutritional balance than traditional maize.

Wheat takes up more of the world's surface than any other crop. Since the 'Green Revolution', the breeding programme of the 1960s and 70s, wheat yields have increased around 15-fold. Scientists bred short plants that converted the sun's energy into grain rather than stem, so preventing the mass starvation in the developing world predicted before the 1960s, at a cost of higher inputs from chemical fertilizers and irrigation. Disease-resistant wheat varieties with high yield potentials are now being produced for a wide range of global, environmental and cultural conditions.

Rice is grown in 113 countries and all continents except Antarctica. Over 90% is grown in the developing world and it feeds about half the world's population.

Potatoes, originally from the Andes, are now grown from the temperate to the tropical regions and produce large yields faster than most of the cereals.

Future Harvest, an educational charity, supports international agricultural research with farmers in developing countries for a world with less poverty, healthier well-nourished people and a better environment. Their research centres across the globe study and improve these crops to provide food security for the growing world population.

).04 The making of garden flowers

Wild flowering plants were brought to our gardens from across the world. Seeds set and grew and new forms arose. Gardeners sped things up by cross-pollinating (hybridizing) promising parents. The offspring, not normally found in nature, are called cultivars. Today old forms have a role in breeding new varieties. Brad Dillon has created a twisty fence to support sweet peas – in the winter its sinuous shapes will remind us of the summer flowers to come.

).05 Colours and dyes

The leaves are green, the flowers are yellow, but the plant growing here (in season) yields a beautiful blue dye - indigo. Joining it at certain times of the year are weld (yellow) and madder (red) to make the triumvirate of best-known British dye plants.

Indigo, which strengthens fibres and heals skin, has been used by blue Ancient Britons, blue-jeaned teenagers, soldiers, sailors and boys in blue. Found in several very different plant species across the world, this blue dye is only formed when the leaves are processed. The dye comes out of the dye-vat yellow and turns blue as it meets the air. Although

natural indigo is still used in many places, synthetic indigo now supplies most of our needs. But new technology means that natural, renewable, non-toxic indigo from European woad plants and subtropical indigo plants, grown elsewhere, can provide work for growers, dyers and designers, and may be able to compete economically with synthetics.

O·06   Eden play

If you are short of play structures why not grow your own? We hope you like these, and we're installing more elsewhere on site.

Play is vital for the young and young at heart. It helps to develop skills and knowledge and connects people with their environment, as well as keeping them fit and healthy.

O·07   Plants for Taste (Biome Link frontage)

Outside the restaurants watch your food grow while you eat and discover stories of some of the herbs we use. The Romans ate parsley to sweeten their breath and discourage intoxication. The ancient Greeks, who were not so keen, fed it to their horses.

Vast quantities of fossil fuels are used to transport food thousands of miles round the world. At Eden we reduce food miles and support the local economy by using local produce in our restaurants whenever we can.

Eden Live

What: Eden Live is our year-round events programme of interactive workshops, talkshops, debates, live music and theatre performances that put champagne in your veins and ideas in your head.

Our free daily programme is led by the Eden guiding and performance teams, joined by our global partners and supporters.

In the spring we bring you Bulb Mania, in the summer Flower Power, followed by our autumn Rainforest Canopy season and finally Winter Tales.

Where: the arena provides a spacious outdoor venue for Eden Live, with performance, music and much else besides. In summer the stage arrives for the Eden Sessions: glorious nights of music and celebration. Near the al fresco café outside the Link you'll find the main (temporary) venue for Eden Live workshops and talkshops. Come and get stuck in.

Ticketed events in support of Eden's stories include lectures, comedy nights, club nights, film shows and, every August, Eden's now famous season of live music events, the Eden Sessions. Last year, Moby, PJ Harvey, Elbow, Badly Drawn Boy and the Thrills played to sell-out audiences. An annual celebration of world music organised by Womad provides the grand finale to the summer. This year promises another world-class programme which will also feature opera and classical music.

Details from the information kiosks around site, in the local press or at **www.edenproject.com**.

0·08 Plants for **Tomorrow's Industries**

We know that plants make food, medicines, provide our oxygen and so on, but they can also act as green factories providing plant plastics, plant bio-composites and plant oils.

Here, we show you the plant factory: the raw materials and some of the products. Imagine a world engineered by, constructed of, and powered by living things. Could plants take over from fossil fuels?

0·09 **Lavender**

Named from the Latin *lavare*, to wash, lavender has a beautiful scent that attracts insects. It has been used to make aromatherapy oils, perfumes, cleaning products, insect repellents, antiseptics and much more.

0·10 Plants and **Pollinators**

Welcome to nature's self-service restaurant. Plants can't move (much), so how do they reproduce? Many do it by

luring insects and other animals to take pollen from one flower to another. The flowers use colour, scent and rewards of nectar and pollen to attract their go-betweens. Insect–flower relationships are often very specific.

Over half our food plants worldwide depend on pollinators, so spare a thought for the insects – your lunch may depend on them. Robert Bradford brought us Bombus the bee.

0·11 Plants for Cornish crops

Traditionally, Cornwall and the Isles of Scilly have been home to early spring crops – bulbs, potatoes, cauliflowers and cut flowers. Times are changing: we can have what we like, whenever we like, flown in from the country round the corner. Increasingly, Cornish growers are supplying local shops and restaurants (such as Eden!), setting up box delivery schemes and growing new types of food crops. Innovative industrial crops such as biomass, oils, herbs, fibre and dyes are also on the agenda.

0·12 Beer and brewing

Beer followed in the footsteps of wheat and barley from the Fertile Crescent, the area which includes modern-day Israel, Lebanon and Syria, and parts of Jordan, Iraq and south-east Turkey. As late as the 1600s many men, women and children in this country drank around 3 litres of weak beer a day rather than risk drinking dirty water.

Reece Ingram, Cornwall-based sculptor, carved our traditional hop poles. Find the ingredients of beer: wheat, barley, yeast and hops, the hop stilt walker, the brewing kettle, the isinglass (the sturgeon's swim bladder) used to clear it – and the magic formula for alcohol.

0·13 Tea

Tea is made from the young leaves of the tea bush, *Camellia sinensis*. Thousands of years ago the Chinese used tea as a medicine. Today there is renewed interest in its health-giving properties. Tea is grown in around 25 countries in the subtropics and the cool, moist, mountainous tropics, from sea level to 2,100 metres – and here at Eden. Jack Everett and Nicholas de St Croix designed our tea-leaf house.

0.14 Education Centre

This is a temporary home for our schools programmes until the new Education Centre is completed (see p. 18). Watch out for would-be explorers heading for the rainforest and the children gathering cake ingredients for our Crazy Chef to cook. Our schools programmes, for all ages and abilities, cover many specific aspects of the curriculum and breathe magic into learning. In the holidays, visitors can get involved too in our public education workshops, part of Eden Live.

0.15 Eco-Engineering

For centuries we have used plants to bind and heal the earth's fragile skin in areas prone to soil erosion and landslips. Traditional knowledge faded as civil engineering techniques took the stage. Today there is a renewed interest in plants, using heavy construction only where it is essential.

0.16 Hemp

To grow hemp at Eden we needed a licence and a physical barrier, so George Fairhurst, professional tall-ship skipper turned sculptor, designed and made this Hemp Fence.

Canvas sails and ship's ropes, tough clothes and bank notes
Oils and cords, insulation boards,
Soap and Bibles, old masters and fibres –

all this from hemp: easy to grow and suited to our climate.

0.17 Hidden crops from the Andes

The Andes is the ancestral home of the potato, which comes in black, orange, dark red, striped, knobbly and smooth forms. Today these potatoes are used in breeding programmes to produce disease-resistant crops, and are being made into crisps and chips, helping to bring income and economic stability to local people.

Other mountain treasures include *arracacha*, *yacon* and *maka*, traditional Andean root crops that can be used to make baby foods, safe sweeteners for diabetics and the Peruvian equivalent of ginseng. They could also be marketed as export crops to help support local Andean farmers on low incomes. Scientists are exploring the usefulness of these crops in other developing countries where it is difficult to grow food because of the high altitude, cold weather and/or high costs of fertilizer.

0.18 Plants for rope and fibre

For centuries people have made cordage, cloth and ropes from strong plant fibres. Without rope we might never have erected the pyramids or Stonehenge, let alone tied up the runner beans. Modern-day uses include geotextiles and non-woven matting for bio-composite panels. Plants at Eden used for fibre production include flax and New Zealand flax. Nettles can also be cropped and give a strong, silky fibre now being used in designer clothes. George Fairhurst created this huge metal giant: a vertical derrick holding multiple lines to form a massive rope.

O.19 Steppe and prairie

The prairies, with waving tall grasses and colourful flowers including Echinacea and Golden Rod, once covered over 1,000,000 square miles (2.5m sq. km.) of the Midwest. They were created by controlled burning – hence the charred timbers shown in our landscape – to attract game and ease travelling. Back on the range, work is underway to restore and re-create them and let the buffalo roam once more.

The pitcher plants growing in the damp area at the front eat insects. Watch your toes!

The Steppe covers vast tracts of Eastern Europe and Central Asia. Many of the wild ancestors of the Founder Crops in the Fertile Crescent such as wild wheat, barley, lentils and chickpeas originated from the moister fringes of woodland steppe.

On the way to the Plants for Fuel exhibit take a look at the Mediterranean Outdoors, where agave, sisal, proteas, the Chilean wine palm and the elephant foot plant all grow on this sheltered, south-facing slope.

O·20 Plants for fuel

David Kemp's flaming industrial plant reminds us of the fossil fuels we've used for centuries. Nip round the back to see what he's planted in the greenhouse.

Plants are living power stations. They turn sunlight into the fuel (carbohydrate) that powers our bodies using carbon dioxide and water as the raw materials. The only by-product is oxygen. We eat carbohydrate, breathe oxygen and produce energy, carbon dioxide and water. It is all a big cycle. When plants are burnt for fuel they release the CO_2 they absorbed during photosynthesis. Problems arise when we burn 'fossil' fuels. The carbon they stored underground for millions of years once again becomes CO_2, increasing levels in the atmosphere, contributing to climate change.

Renewable plant fuels such as wood (used worldwide), energy crops (e.g. willow coppice), agricultural wastes (e.g. straw) and biofuels (e.g. biodiesel from rapeseed) are all carbon neutral: if you replant what you harvest the CO_2 levels in the air will remain about the same. But how much land would be needed for everyone? Other energy choices include harnessing wind and water, the hot rocks beneath us, or copying the green plant and going for solar power. At Eden we buy all our electricity from wind farms. You can do the same.

O·21 Plants in myth and folklore

Myth, folklore, stories and poems keep plants alive in our memory. If remembered and revered they have a far better chance of staying alive in the ground too. Eden performers tell their stories here and pop up across the whole site. Everywhere, everyone has his or her own Green Man.

Pete Hill and Kate Munro created our magical Story Pavilion. Our maze is the same as the ones carved into the granite near Tintagel, and laid out in stones on St Agnes, Isles of Scilly.

0.22 Biodiversity and Cornwall

Biodiversity – the vast variety of life – makes up an intricate, interdependent web. Living creatures have evolved into many different forms that live together and share resources. About 1.75 million species have been named, but it's estimated that there may be over 10 million. Each one plays a part in the great web.

We highlight biodiversity in many of our exhibits but the issues surrounding biodiversity conservation are true of habitats and ecosystems worldwide. Why conserve biodiversity? It's a matter of survival: the earth and its plants and animals keep us alive. Research also confirms our intuition that the 'wild' places help us feel good – and what right do we have to destroy it anyway?

Cornwall has many habitats, but from heathland to hedge people have shaped almost all of them. So come and take a trek round Wild Cornwall and discover some of the conservation work being done in partnership with English Nature.

Atlantic woodland: the south-west-coast Atlantic woodlands are among the least disturbed of our semi-natural habitats. They contain native oak, willow, ash and hazel, pruned by the wind, dwarfed by the poor soil, and clothed in ferns, mosses and lichens thanks to the clean air. There is a rich diversity in these temperate rainforests, but many species are small, green and easily overlooked. English Nature is recreating 500 to 1,000 hectares of woodland landscape in the Cornish clay area, restoring a habitat lost about 100 years ago.

Kate Munro and our Green Team crafted the metal wind-pruned trees, so typical of Cornish uplands. Chris Drury created the Cloud Chamber. Whatever is passing in the sky above will be projected in here beneath your feet. Magic.

Cornish heathland: lowland heath is a rarer habitat than tropical rainforest. Since 1945 Cornwall has lost more than 60% of its heathland. This landscape of dwarf shrubs and grasses was created 4,500 years ago when woodland was cleared for grazing, hunting and agriculture. Continued grazing, burning and management maintain the landscape and prevent it reverting to scrub and woodland. Conservationists are even restoring vast areas of wasteland, such as china clay tips, to their former glory, and the heathland is being used once more for recreation, free-range meat and heather beer.

Farmland habitats: three-quarters of the land in Cornwall is farmed. Responsible farming provides rich habitats. Unsprayed field margins and Cornish hedges (mini-mountains of soil and rock) provide homes for hundreds of plants and animals. Ancient granite hedges enclose small fields in Penwith, west Cornwall, and these are believed to be some of the oldest in the world, dating from the early Neolithic, 4,500 BC.

Endangered species: the early gentian *Gentianella anglica* is rare and beautiful. The lichen *Heterodermia isidiophora* is also pretty rare, but not very pretty. English Nature, Plantlife and others are working to restore and recreate habitats for these endangered plants. The Cornwall Biodiversity Initiative (CBI) unites local conservation groups. We stock CBI products in our shop to support this work.

Peter Martin and Sarah Stewart-Smith, Cornish stone sculptors, have helped to immortalize some of our rare species in stone.

0·23 **Copper** connects

As well as harvesting plants, we mine rocks and minerals to use in our everyday lives. Mining is a controversial but essential industry that we need to manage better as part of our stewardship of the planet. This exhibit, in construction in 2004, starts the story.

0.24 Big Build 2

The Big Build 2 exhibit is across the bridge from the Visitor Centre. At Eden we always intended to have three covered Biomes, an inspirational Education Centre and assistance to help you out of the pit. However, by 1998 we had only raised enough money either to build a scaled-down Eden or do part of it as intended. Fingers crossed, we plumped for the latter and built the Humid Tropics and Warm Temperate Biomes – and we built them BIG. Millions came and now with several years' experience and thanks to further funding from the Millennium Commission, the South West Regional Development Agency, the European Regional Development Fund (via Objective One), Restormel Borough Council and Viridor we have got out the diggers to take Eden to the next stage.

On the shopping list:

One **Education Resource Centre**. A **smooth arrival** for visitors. One **Waste Recycling Unit**. **Facilities** to take care of the staff who take care of you. **Escalators**, and other ways, to get you in and out of the pit. One **Gateway Building**. One **Green Room Café**. **Covered routes** and some **gardening** work.

Our new Education Centre, ready in 2005, will be an inspirational hub for events, exhibitions and learning ... for all. An exhibit in its own right, the building has been inspired by natural form and will be crafted from natural materials.

In our Waste Recycling Compound, being constructed in Pineapple car park, you will be able to discover how we recycle paper, glass, metal, cardboard, wood and wood waste and, in our biodigestion unit, organic waste.

We are currently planning and fundraising for our final covered Biome. The Dry Tropics Biome will bring stories of survival and diversity – of people and of plants that live on the edge – in the hard, waterless parts of the world.

Following on Temporary displays, trails and new exhibits

Eden is a project, ever changing, ever evolving. Many of the plants we grow, especially in the Outdoor and Warm Temperate Biomes, are seasonal so we use temporary displays to keep our stories alive while the ground and plants are being prepared for the next year. Bulb Mania 2004 is our first bulb festival, from March to May; 300,000 bulbs will paint the landscape, celebrating Cornwall as one of the world's principal daffodil providers. Look out for winter displays and guest exhibits by local nurseries in the Visitor Centre.

H·01 Introduction to the **Humid Tropics**

This Biome, the largest conservatory in the world, is 240 m long, 110 m wide and 50 m high. It contains over 1,000 plant species, which have shot up since they were planted (from September 2000 on), and we've lost less than 5%. See if you can spot the birds, lizards and frogs, here to help control the pests (p.56). The moist air is kept between 18° and 35°C. There is an exit point at the Malaysian House (H.03), a cool room in West Africa (H.04) for emergencies, and seats and water fountains to keep you comfortable on your way round.

Short, rough guide to tropical rainforests

Jungle trees race up to the light, some growing several metres a year. Climbers hitch a ride and epiphytes, such as orchids and ferns, live high in the living skyscrapers. Different rainforest species share certain characteristics because they have evolved to cope with the hot, steamy conditions. Look out for large, shiny leaves with 'gutters' and 'drip' tips designed to shed excess water, and, when they have had time to grow, trees with stilts or buttress roots for support in the thin soil. Spot the plant leaves with purple undersides that act as reflectors to bounce back the 2% of light that filters through to the forest floor, so getting a double dose.

Rainforests are a huge resource and contain thousands of different plant species. Some, such as cocoa and rubber, have been bred as crops. New crops will undoubtedly be discovered. Fewer than 5% of the plants in the Amazon, for example, have been tested for their medicinal properties.

But conserving forests isn't just about future crops. These massive, living ecosystems are a vital global environmental resource. They are the largest store of carbon on earth, sucking in carbon dioxide from the air as they grow. They also make rain. Water travels up inside the trees, evaporates, makes clouds, rains back down, and is taken back up by the trees. Changes in areas of rainforest therefore lead to climate change; whether it gets hotter, colder, dryer or wetter is still under debate, but we do know that local changes have global consequences. Rainforests are cleared for agriculture, mining, development and timber, but they can also re-grow or be replanted and managed sustainably for the future.

So join us on a trip through the world's rainforests, discover how people, near and far, rely on the plants from the tropics and how they are managing the land to meet their needs and conserve the environment.

Look at the crates of products on the *Tropics Trader*, the huge ship bow created by David Kemp (of Industrial Plant fame, see p.15). See anything familiar? The products we need and want, past and present, have far-reaching effects. What you buy, what you say, how you act, can all make a difference.

People have always used and worked with plants. On your right, just after the *Tropics Trader*, is papyrus, once used by the Egyptians to make parchment, and the traveller's palm, which reputedly grows on an east/west axis in the tropics and holds fresh water in the base of the leaf stalks – helpful to thirsty travellers who may have lost their bearings.

Introduction to
  the Humid Tropics   H.01
Tropical islands   H.02
Malaysia   H.03
West Africa   H.04
Tropical South America   H.05
Crops and cultivation   H.06
Cola   H.07
Chewing gum   H.08
Rubber   H.09
Timber   H.10
Cocoa and Chocolate   H.11

Palms   H.12
Rice   H.13
Coffee   H.14
Tropical displays   H.15
Sugar   H.16
Mangoes   H.17
Bananas   H.18
Tropical fruits   H.19
Bamboo   H.20
Pineapples   H.21
Pharmaceuticals
  from the land   H.22
Spices   H.23
Cashews   H.24
Tropical dyes   H.25

oint

H.07 H.08 H.09 H.13
H.06 H.10
H.05 H.11 H.12 H.15 H.16
H.20 H.14 H.17
H.21
H.02 H.22 H.19 H.18
H.03 H.23 H.24
H.25
H.01
H.04

Early Exit

Biome Entrance

H·02 Tropical islands: conserving the land

Note the mangroves down on the left as you skirt the lake. Mangrove swamps link land and sea, protect the coast and provide fuel, timber and a habitat for fish.

Islands are home to unusual plants and animals that have developed in these isolated environments. Some are relics of a lost world, extinct everywhere else; others have evolved into strange forms, such as giants. All are irreplaceable. Island communities are isolated too, with few resources to support their global responsibility for biodiversity conservation. Climate changes, invasion of aggressive species and human settlement pose serious threats, but many countries now have conservation programmes that offer hope.

Protected species include the rare and extraordinary Coco-de-Mer from the Seychelles. Its seeds, the largest in the world, look like giant bottoms, which has led to overharvesting as trophies and for their perceived aphrodisiac qualities. Unlike coconuts, they die if they travel in salt water to other islands. Every seed that grows is registered and numbered. The Seychelles government kindly donated two seeds, one of which is planted but may take years to show above ground. The other you can see in our peepshow exhibit.

The rare white-flowered *Impatiens gordonii*, a relation of our Busy Lizzie, is found only on two islands in the Seychelles and is under threat due to competition from introduced species and loss of habitat. Eden is working with the Seychelles government and the University of Reading to propagate it and safeguard its future.

Some of the offspring can be seen on the right-hand slopes after the bend.

H·03 Malaysia:
Orang dan Kebun (people and garden)

After passing the Malaysian map and the bamboo you enter the mainland rainforests of Asia. In front of the house is a contemporary home garden which provides a year-round food supply.

The Royal Society S.E. Asia Rainforest Research Programme helped Eden with this project. The garden was based on smallholdings at Kampong Tampinau, a village in Sabah, Malaysia, and the house built by Mark Biddle, Hamish Thomson, Brian Lloyd and the Eden team.

The garden is zoned – herbs and flowers nearest the house, then vegetables, then fruit and other useful trees further out. The miracle or horseradish tree, *Moringa oleifera*, has edible leaves, beans, flowers and roots. Beside it stands the neem, known throughout the east as the world's most useful tree, providing medicine, fuel and food. There are parallels with our own gardens. Winged beans replace runner beans; both help to fertilize the soil. Pak choi, taro and rice replace cabbage, carrots and potatoes. The garden also provides building materials, medicines and produce to barter

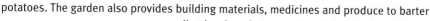
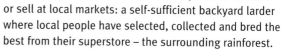

or sell at local markets: a self-sufficient backyard larder where local people have selected, collected and bred the best from their superstore – the surrounding rainforest.

Behind the rice paddy on the left of the path we are creating a secondary forest where the house dwellers would gather their fruit; and on the other side of the stream, primary forest. Beyond the Malaysian garden where the path widens on the corner you will find a stunning Bo tree with delicate pointed leaves. It was under such a tree that Buddha meditated to find enlightenment.

Continue past the rushing waterfall and up into West Africa.

H·04 West Africa: managing the land

How do you feed the soil, feed yourselves and replant the forest simultaneously? The Eden team journeyed to Cameroon to find out from the local people in and around Ndoumdjom village, an agroforestry region in the Adamaoua province of Cameroon, with help from RECOFON (a Cameroon NGO).

On your right is the *taungya* system: crops like coffee and cacao that need shade are grown beneath useful trees such as *Prunus africana*, used to treat prostate cancer,

in a tiered production system. As the trees grow bigger the garden once more becomes a productive crop forest, good for the local people, the economy and the land.

Round the corner and up on the left-hand slope, alley cropping shows how to grow food and prevent soil erosion on steep land. Trees are planted along the contours between maize, sorghum and other crops to bind the soil, increase soil fertility (if they are legumes or if branches are used as a mulch) and often produce a crop themselves. Vetiver grass, also used to stabilize the land, creates less shade than trees. Originally from India, it has been used for 3,000 years as a perfume, insect repellent and boundary hedge.

The Eden team, aided both by colleagues in Cameroon and by Cornish archaeologist Jacqui Wood, recreated the African rondavel in the forest garden, which provides temporary accommodation for farmers whilst tending cattle and/or crops. It contains the bare essentials: shelter, a sleeping area, a fireplace and a storage place.

Past the rondavel the path splits. The high road, which takes you past the waterfall, provides excellent views but can get crowded and there are steep steps down at the other end. The low road is flat and takes you past the shifting cultivation exhibit. The paths meet by the lily pond. Above the lower path on the left, our kapok tree: one of the tallest trees we planted in the Biome and still growing.

H·05 Tropical South America:
shifting cultivation and plant-gathering

Whatever route you choose, once you're past the map you've crossed the Atlantic. The Eden team visited Guyana, where the local people and the Environmental Protection Agency taught us about their forests and gardens. Some villagers practise shifting cultivation, moving their garden on to a new plot every year, rotating round the village on a 14-year cycle. They cut about an acre of forest, let it dry and then burn it to provide fertility from the ashes. Staple crops such as cassava and sweet potatoes are then planted for most of the year, as seen on the right of the lower path below the lily pool. Cassava often contains prussic acid (hydrocyanide), a poison that has to be washed out before cooking. After harvesting the land is left, and the forest soon reclaims it. The forest also provides a natural garden. Plants are collected for food, fuel, medicine and materials.

Francisco Montes Shuna and Yolanda Panduro Baneo, inspirational shamanic artists from Peru, painted the cliff face above the upper path with pictures of the spirit lives of rainforest plants. They showed us how myths and stories help to keep the knowledge alive in local cultures and the plants alive in their forests. The project was supported by the Arts Council and the October Gallery.

From the clearing at the base of the steps, look out for a large water chestnut tree and a dugout canoe, useful on the South American waterways.

H·06 Introduction to crops and cultivation

Every day rainforests touch our lives, and we touch them. Rubber gloves, oils in processed foods, chocolate, lip gloss and food colouring can all come from the tropics. Many of the things that we do and buy affect the future of the rainforest.

Paul and Will, of Dead Cat fame, brought us the arch where products (cola drinks, cupcakes and chewing gum) occasionally meet their makers.

Our first stop: a boat-shaped bed where we would like you to meet two unfamiliar plants whose products you probably see every day.

H·07 Cola

This is probably the most-used Latin name in the world. Cola, an African tree with caffeine-rich seeds, is part of the age-old culture of West Africa. Cola, a sparkling flavoured drink, is part of a new global culture.

H·08 Chewing gum

Chicle, a milky latex harvested from the sapodilla tree, *Manilkara zapota*, can be made into chewing gum. This doesn't harm the tree (if it is not overtapped), provides a living for local people and makes gum that can clean your teeth. The trees also yield tasty fruit that makes delicious ice cream.

H·09 **Rubber**

Take a rest on our tyres by these elegant trees. Rubber trees, *Hevea brasiliensis*, have been tapped for their milky latex for centuries in tropical South America to make rubber boots, bottles and balls.

In the 18th century European scientists stepped in with waterproof clothing and catheters. Then came the car, and a spiralling demand for rubber.

A frantic global search for plants that produced alternatives was followed by the rise of a new industry in Asia, cultivated rubber. More demand, more supply, oversupply and complex schemes to restrict output caused rivalry; wars restricted supplies and introduced synthetic rubber. Rises in petrol prices, AIDS and the spiralling demand for condoms and rubber gloves, let natural rubber bounce back. Today, some of this rubber is being sourced from smallholdings and from designated areas of rainforest rather than plantations.

As to the future, researchers are working on producing albumin from rubber tree latex. This is a protein in human blood given in transfusions. The by-product this time? Rubber for aircraft tyres.

H·10 Tropical **timber**

Pop down the steps to take a look at tropical timber. Here is a range of forest trees, teak, baywood and thingham, used for their wood.

The rainforests are getting smaller. What can we do? We can choose to buy sustainably produced timber from a certified source such as the Forest Stewardship Council. This only tackles the tip of the iceberg in the tropics; logging profits aren't huge and many contracts are short-term, so companies aren't willing to invest in long-term sustainability. There is also a big threat from the domestic use of tropical timber for construction, firewood, crafts and agriculture. More damage can be caused to the surrounding forest by dragging the tree out than the loss of the tree itself.

Ways forward include research into using different timber species; involving forest dwellers in the decision-making process; and developing sustainable livelihoods from non-timber forest products that can be harvested without harming the forest. These products are beginning to appear in our shops.

H·11 Cocoa and Chocolate

Look for the tiny white flowers on the tree trunks. These, after being pollinated by midges, grow into yellow pods containing cocoa beans. From the bottom path, see our 'Mayan tapestry' cartoon strip, a brief history of cocoa.

Cocoa beans, brewed with chillies, were drunk by Mayan and Aztec nobles. The Latin name, *Theobroma cacao*, means food or drink of the gods. Conquistadors brought cocoa to Europe, where it was later made into sweet chocolate. First reserved for the rich, chocolate went on to become part of the emergency rations kit for armies and rescue teams. Now it's a 'luxury' that nearly all of us can have. In the UK that means we eat around 10 kg of chocolate each a year.

Most of our cocoa is grown on smallholdings in West Africa. Cocoa disease has hit some of these farms. Rather than use chemicals or cut down rainforest and plant on clean land, scientists are working with the BCCCA (Biscuit, Cake, Chocolate and Confectionery Association) and the global chocolate and cocoa industry to marry West African cocoa trees with their wild – and disease-resistant – ancestors from the South American rainforest. The resulting disease-resistant plants can be used on existing smallholdings.

Around 2.5 million farmers grow cocoa, relying on the income from our enjoyment of chocolate. In Ghana, the Government fixes the price per kilo each year so it's the same for all farmers. In countries where the market has been liberalized, cocoa farmers receive prices that can change daily in line with the London and New York cocoa markets. The UK chocolate industry supports a number of schemes to improve the

livelihoods of communities in cocoa-growing areas and protect them from price fluctuations. Fairtrade is one approach. Fairtrade provides producers with a stable price which covers their production costs, along with a premium that their organization can reinvest either in the business or in social and environmental schemes among the wider community.

H.12 Palms

Back on the main path, the palms. In the tropics a huge range of palms are used by local people. Stems, leaves, trunks, sap and fruits provide walls, thatch, ropes, boats, sago, sugar, cooking oil and much more. Coconuts 'wash up' on European shores, used mainly for flesh (copra) and fibre (coir). We may be more familiar with piña coladas, hair conditioners and doormats. However, on the international market the oil palm reigns supreme. Palm oil is found in many of our processed foods, cleaning products and cosmetics. Supply chases demand and plantations march into the rainforest. Plantation work is hard and dirty, but people work dirty and hard, aspiring to, but rarely getting, the world on the satellite TV. 'Don't cut down the forest,' we say. Who are we to talk? We already have our TV dream.

Where next? New sustainable initiatives are slowly emerging: planting oil palms on degraded land rather than newly felled virgin rainforest; new co-operatives to enable workers to control their own destinies. We are working with partners such as the Natural Resources Institute, Rainforest Concern and others.

H.13 Rice

Boiled or fried, pudding, paella, risotto, sushi ... how often do you eat it? 2004 is the International Year of Rice.

Rice for life: this grain has fed more people for longer than any other crop, and today nourishes around half the world. Rice is deeply respected in many cultures. We see a man in the moon. In Vietnam they look at the moon and see the Rice Goddess, stacking her freshly harvested rice in the shade of a Bo tree. Cornish artists Phil Booth and Louise Thorn worked with Japanese rice-straw sculptors Eio Okumura and Professor Yoshihisa Fujita to make our Shimenawa. Traditionally this was made after the Japanese rice harvest to celebrate the Shinto gods. The Shinto religion pays great respect to nature, promoting harmony between humans, plants, animals and the landscape. The Rice Goddess watches over the many rice landraces (ancient types of crop plants, whose genetic diversity helps them adapt to their growing environment), one or more of which may be the parent of the 'rice of the future'.

In the 1960s Future Harvest's International Rice Research Institute (IRRI) was set up in the Philippines and in the next thirty years the global rice harvest doubled.

Different rice landraces were crossed to produce high-yielding semi-dwarf varieties. IR8, released in 1966, was one of the first 'miracle rices'. An Indian farmer called Ganesan was so impressed with his bumper IR8 rice harvest that he named his son after it. The increased income from the crop later enabled Ganesan to send IR8 (IR-ettu in Tamil) to college.

IRRI scientists are looking to some of the hundred thousand rice landraces for assistance in future breeding programmes.

Scientists have recently crossed Asian and African rices to create 'New Rice for Africa' to bring hope to the hungry.

H·14 Coffee

Coffee is not only big business; it has been a driving force in history. Starting life in Ethiopia, coffee travelled to the Yemen, took a pilgrimage to Mecca, wound up the whirling dervishes and gave birth to the coffee house in the Middle East. Exchanging news and views, wheeling and dealing, chin-wagging, and even plotting – these cafés provided the place, and the coffee the stimulation. By the 1600s coffee and its houses reached Britain and continued to spawn intellect and commerce. Lloyds of London, the *Tatler* and the Royal Society all started life in coffee houses.

Coffee fuelled the industrial age, and today is probably the most valuable tropical product on the world market, amounting to an estimated US$70 billion of retail sales per year, only $5 billion of which is earned by the exporting countries. At the beginning of the chain, things are often less rosy. Many beans are still picked by hand, labour is high and income low. Eden are working with partners and supporters to bring you the news of how producers and organizations are working to provide a sustainable coffee future.

We are developing Eden-brand coffee, using Rainforest Alliance-certified beans, which will be served in our restaurants and sold in our shop. Our staff have also harvested and tasted a cup of coffee made from beans from the Biome.

H·15 Tropical displays

Depending on the time of year you will find a range of tropical displays including fruits, vegetables and flowers.

H·16 Sugar

Some of our sugar comes from sugar beet, grown in temperate climates, and some from sugar cane, a huge tropical grass. Cane probably first satisfied people's taste for sweetness around 10,000 years ago in New Guinea, its original home. By the 14th century it had become a luxury spice in Europe. Today, in the UK, each of us consumes around 35kg every year. In order to satisfy our demands the world harvests more raw sugar cane than it does wheat!

What does this mean for the sugar producers? Sugar's rise to stardom has been tainted by human exploitation and trade wars. However, it has also brought financial security, and social benefits such as schools and hospitals, to many developing countries, especially those supported by preferential trade agreements which act as 'trade aid'. Today, sugar prices are down and trade rules are changing. What of the future? Improved techniques have been introduced in recent years, including some organic production. Using sugar as a fuel could also help the industry and the climate. The impact of trade rules needs to be considered. Equitably traded sugar could bring social benefits.

H·17 Mangoes

One of the world's most important tropical fruits, mangoes can be eaten savoury, sweet, pickled or salted and in some countries are a vital famine food. The flesh makes medicines and wine; the nut oil, cosmetics; and the wood, traditional drums and furniture.

Mangoes have been grown in India for about 5,000 years and come in many shapes, sizes and colours. Trees are planted to liberate the souls of Hindu ancestors, the fruits symbolize immortality and love and have been hailed an aphrodisiac by some.

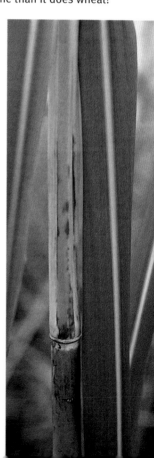

H.18 Bananas

Bananas come in many shapes, colours, sizes and guises. Over 80% of them are used in the tropics for sweet and savoury foods, beer, cloth, roofing material and more. In Africa 70 million people eat banana and plantain every day as part of their staple diet.

Less than 15% of the world's bananas, mainly Dwarf Cavendish, are exported. People didn't always like them and comedy was used as a marketing tool: 'Time flies like an arrow, fruit flies like a banana,' as Groucho Marx said.

Now bananas are on the bestseller list. We eat around 130 each a year. Europe has always supported the traditional small farms in the African, Caribbean and Pacific (ACP) countries which rely heavily on banana exports to survive. This led to the 'Banana Wars', a dispute which brought a ruling from the World Trade Organization that Europe should reduce its economic support. The upshot tended to favour larger farms which can produce cheaper fruit, though often at the expense of incomes and working conditions. Fairly traded banana schemes that ensure equitable returns for the farmers have been introduced in Latin America and Caribbean countries such as the Windward Islands. Check your bananas!

Save the banana: Cavendish bananas may be under threat from Panama disease, so scientists are looking to breed disease-resistant types. The Future Harvest Centre, INIBAP (International Network for the Improvement of Banana and Plantain), has a collection of over 1,000 different types held 'in trust' for the public good which may be used in breeding programmes. They are working to deep freeze (cryopreserve) this precious plant material for future use. Why? Because most plant 'gene' banks are stored as seeds – and most bananas don't have any.

H.19 Tropical fruits

Soursop, custard apple, breadfruit, jackfruit, akee, longan, Surinam cherry and Jaboticaba are some of the latest fruits to hit the EU/US market. What are the implications of our desire for unusual tastes? These niche tropical fruits could provide an income for the smallholder and help to maintain rural communities.

Co-ops and export associations can help the small farmers work together on issues such as EU legislation on traceability and pesticide residue levels and help prevent them being squeezed out of the market by the big players. Systems that take account of the environment and social issues could lead to a sustainable fruitures market!

H·20 Bamboo

This green gold of the east is used by half the world's people. It makes homes and furniture, food and fuel, music and medicine, paper and poles, scaffolding and suspension bridges. Its hollow tubes are strong but light. Within its tissues short, tough fibres sit in a resilient matrix, providing nature's version of fibreglass.

Housings and Hazards, who work to make affordable, hazard-resistant housing available to vulnerable rural communities around the world, led the construction of our bamboo house frame based on a design originally created by renowned Colombian architect Simon Velez.

Grow your own house in five years: fast-growing bamboos are ideal materials for low-cost, low-impact, earthquake-resistant houses. This renewable resource provides materials and employment and unites science and art, rich and poor, high tech and low tech, city and country – a real bridge-builder.

H·21 Pineapples

Pineapples don't grow on trees! They grow on stalks from spiny rosettes of leaves. The fertilized flowers form fleshy fruits that fuse to become a pineapple. A must-have at Victorian dinner parties, pineapples were nurtured in their hot houses. Heligan Gardens, just down the road from Eden, grow them in this style today. Today we all 'must have' and so pineapples tend to be produced on large farms hundreds of hectares in size. Crops grown on the same plot each year may use significant amounts of fertilizers and pesticides, so scientists are looking to Integrated Crop Management, organic cultivation techniques and GM technology to reduce or remove the use of chemicals. The fairly traded pineapple is on the increase too.

Pineapples produce alcohol, pina cloth which is finer than silk, candles, animal feed and medicines as well as chunks and rings. Pineapples take 1.5 to 4 years to fruit. Grow your own by cutting (or screwing) the top off a fruit and planting it in a pot in a warm place.

H·22 Pharmaceuticals from the land

Conserving the rainforest's medicine chest may provide cures for today's illnesses. Plants such as Madagascar periwinkle, source of the alkaloids used in treating leukaemias and Hodgkin's lymphoma, are grown as a field crop since the alkaloids they contain are too difficult or expensive to synthesize.

Drugs are often discovered by good fortune rather than mass plant screening. The Madagascar periwinkle was used locally as a folk remedy for diabetes. Research into its effectiveness showed it was no use for diabetes but did have cytotoxic activity. It takes 2 tonnes of leaves to extract 1 gram of vincristine, the alkaloid used in chemotherapy, which will treat one child for 6 weeks.

H·23 **Spices**

Squeeze and sniff. After absorbing the mood and scent of tropical spices see if you can solve the riddles on the spice boat created by local (but internationally renowned) artists Bill Mitchell and Dave Mynne. The answers lie somewhere on the drawers!

Today spices are cheap. In the past they were worth their weight in gold and shaped the world as we know it. Nutmeg was thought to cure the bubonic plague. Yet in the mid-1300s it was along the spice route from Central Asia that the Black Death first travelled to Europe. It killed a third of the population in five years.

The Arabs monopolized the land-based spice trade to the West until the 15th century. Western Europeans were spun yarns to keep their traders away – of Arabs fishing for spices by moonlight, of cinnamon harvested from the nests of ferocious birds and of boiling seas. Of course in the early days they also thought the world was flat!

H·24 Cashews

These fast-growing, drought-tolerant trees produce nuts after 3 years and can live for 50. Seeds, unusually suspended below the fruit, provide the highly prized nuts, and the shells provide cashew-nut shell liquor (CNSL).

Why are cashews more expensive than peanuts? Because roasting, shelling and cleaning the kernels is a delicate and laborious process and CNSL is a highly corrosive substance, exploited traditionally in the treatment of ringworm and warts. Globally, CNSL crops up in marine paints, heatproof enamels, brake pads and more recently as a resin in completely plant-based 'ecocomposites'. Traditionally bio-composites are made from plant fibres, such as hemp, embedded into fossil-fuel-based resins.

H·25 Tropical dyes

Under the skin of our cultural diversity we have much in common – plants provide the coloured backdrop to our lives. Bark, leaves, roots and flowers dye our skin, food, clothes and hair. Brazil wood was originally a name given to sappan wood, from Asia. Why? Because it means red-hot coals – the colour of the tree's dye. When the Portuguese landed in South America (around 1500) they found other red dye trees and named the land Brazil after the dye wood they already knew.

The seeds from the South American tree *Bixa orellana* (annatto) provide a red dye used locally as a body paint, insect repellent and hair dye. We use this tasteless dye from plantation-grown trees as a food colouring in sweets and cheese.

Following on   Temporary displays and trails

Look out for our mobile interpretation stations and market stalls where guides and storytellers speak their wares, brim with stories, question-and-answer sessions and occasionally offer tasty samples of tropical wares.

You may also spot the Flybot (*left*), a computer-controlled helium-filled balloon with a mounted camera for science, education and entertainment. This will be a central point of our Canopy Programme, part of Eden Live's 'New Ways of Seeing' event, in autumn 2004.

The Warm Temperate Biome

W.01 Introduction to the Warm Temperate regions
W.02 The Mediterranean Basin
W.03 South Africa
W.04 California
W.05 Introduction to Crops and Cultivation
W.06 Fruits of the Mediterranean
W.07 Cork
W.08 Tobacco
W.09 Vegetables and herbs
W.10 Citrus
W.11 Grape vines
W.12 Peppers
W.13 Grains
W.14 Pulses
W.15 Sunflowers
W.16 Tomatoes
W.17 Olives
W.18 Perfume

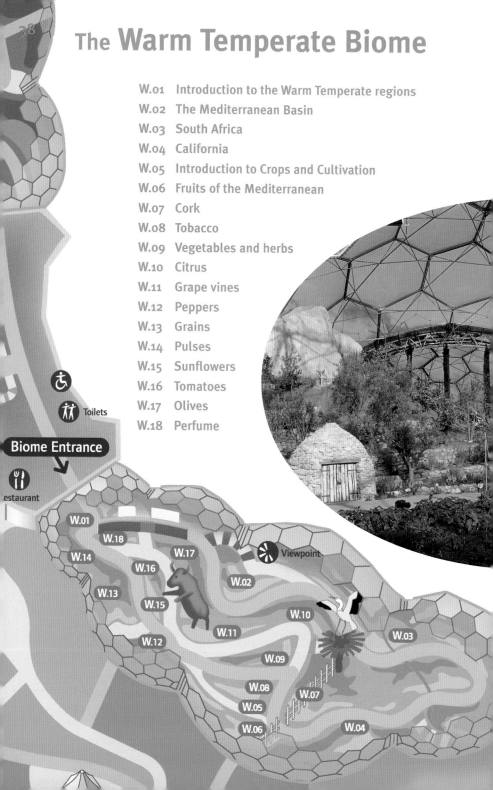

Toilets

Biome Entrance

Restaurant

W.01
W.18
W.14
W.17
W.16
W.13
W.02
Viewpoint
W.15
W.11
W.10
W.12
W.09
W.03
W.08
W.07
W.05
W.06
W.04

N·01 Warm Temperate regions: an introduction

Halfway between our wet green woods and the world's deserts are the warm temperate Mediterranean regions, characterized by hot dry summers and cool wet winters. They are found on the western sides of continents between 30–40° N or S latitude, and are caused by cold ocean currents, trade winds and topography. They include parts of California and South Africa, S.W. Australia and Chile as well as the Med itself – the warm, sunny holiday places. Natural gardens bloom in a gardener's nightmare of drought, scorching sun, poor, thin soils and fire. Plants have spines, waxy evergreen leaves or small, grey, hairy leaves, all of which serve to protect. Shrubs are more common than trees, bulbs hide below the summer-scorched soil, and annuals bloom in a riot of colour after winter rains.

In this Biome the air is kept between 15 and 25°C in the summer and a minimum of 9°C in the winter. Breathe in ... the scent comes from the plants' protective oils. These may act as bug repellents and vapour barriers to reduce water loss. Our Biomes, unlike glass, transmit UV light, which also increases plant oils – so on sunny days don't forget to protect yourself too.

Conservation and cultivation

The plants are tough but their environments are fragile. Intensive grazing has caused soil erosion, imported plants have threatened native species and land has been developed. We take you on a journey through the 'wild' landscapes of the Med, South Africa and California (many of which have been shaped by mankind through the centuries) and on to the intensively cultivated areas where water, food and shade have created a kitchen garden for the supermarkets of the world. Cornucopia?

W·02 The Mediterranean Basin

When you don't cultivate the land in the Mediterranean the land dies – Fernand Braudel

Just as you enter the Mediterranean area the path splits. The upper route, with steps, takes you to the Outlook, where you get a good view across the whole Biome. The lower route, following the golden mosaic path, has no steps. The paths meet before you enter South Africa.

Elaine Goodwin, mosaic artist, created the Liquid Gold pathway as a celebration of the long tradition of olive oil as a symbol of light, life and divinity. Look out for the subtle images of doves – one for each Mediterranean nation.

Culture's cradle: natural Mediterranean vegetation has been cut for timber and firewood and cleared to plant crops for hundreds of thousands of years. One third of Crete is terraced – witness to the long period of human effort to grow food here. Without the olive and the vine, Mediterranean civilization might never have begun. The 'natural' landscape we see today is the product of both nature and mankind.

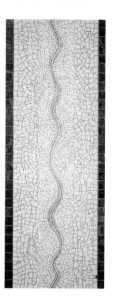

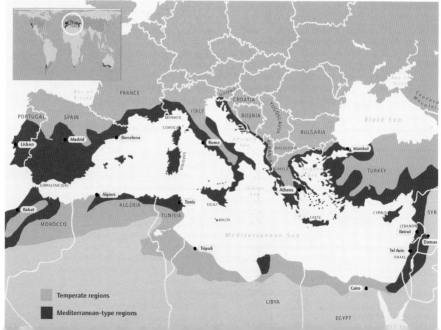

Temperate regions

Mediterranean-type regions

Classical wisdom: 'The Earth conceives and yields her harvest ... but if anything goes wrong, it is not deity we should blame, but humanity, who have not ordered their lives correctly' – Plato

Maquis and Garrigue: the French underground movement in World War II was called the Maquis, because they hid out in this habitat with its prickly oak, juniper, wild olive, laurel, myrtle, tree heather and broom. Partially man-made, maquis can grow into woods if not grazed. Garrigue, with shorter plants and more herbs, is found where there is less soil moisture. Both habitats contain unique plants, insects and reptiles, but can get overlooked, having no spectacular birds or mammals. Do we only save the pretty things?

Traditional olive groves: the ancient terraced olive grove supports far more animal species (insects, reptiles, birds, bats, etc.) than a pine forest. However, people are leaving mountain farms for work on the coast. The goats remain, grazing and knocking down the terraces. One solution proposes focused subsidies, managed locally, to sustain traditional crops and farming. Others argue that subsidies themselves are at the heart of the problem. Buying traditional foods and natural products, seeking out quality and taste, farm holidays: all can help conserve these fragile environments and communities.

W·03 South Africa

Cross the equator in just one step to the unusual plants and wonderful scents of South Africa. The map, which can be found where the path bends round to the right, will help to put the regions described below in context.

The Fynbos, botanical hotspot: the Cape Floral Kingdom has the richest density of different plant species on earth, globally important and very fragile. Fynbos, with around 7,000 plant species (5,000 of which occur nowhere else in the world), covers 80% of this Kingdom, stretching for 46,000 sq. km. either side of the Cape, in a band no more than 200 km. from the southern coast. 'Fain-boss' is Afrikaans for 'fine bush' and refers to the evergreen, fire-prone shrubs that live in this nutrient-poor soil. Because we made our own soil at Eden we were able to create a nutrient-poor mix for this area.

Formed millions of years ago from the ashes of drought-stressed forests, the Fynbos has been extensively fire-managed for conservation since the 1960s. Plant groups include restioids (rush-like plants), proteoids (including many beautiful Proteas), heather-like plants and stunning lilies, orchids and irises. The Fynbos is threatened by

urban spread and development, uncontrolled fire, agricultural conversion and invasive alien tree species.

Fauna and Flora International (FFI), one of our partners, purchased 1350 hectares of Fynbos, Flower Valley, between 1999 and 2002, saving it from being ploughed up to grow vines. They then established sustainable harvesting methods for cut flowers, influencing the management and conservation of Fynbos over an area of 25,000 hectares on surrounding farms. Profits from cut flower sales are invested in conservation, education, business and research.

Little Karoo: this area, with its muted grey foliage, gives our South African section its scent. In reality the Little Karoo is a long, narrow group of semi-arid valleys behind the southernmost coastal mountain range of South Africa. It bakes to 50°C in the summer, freezes in winter, and droughts are the rule. Today much of the valley is irrigated for crops, but the surrounding hills house ice plants, aloes and many types of daisy. The Little Karoo is famous for its ostriches, good wines and spectacular scenery.

Namaqualand: the red desert, about 250 miles north of Cape Town, blooms into a multi-coloured carpet after winter rain. The seeds are temperature-specific, so different flowers germinate in different years depending at what time the rains come. They take it in turns to share out the scarce water. Many plants in these regions are at risk of extinction, yet you will find their descendants, including geraniums and daisies, in your garden.

W.04 California

By the Harley motorbike, step back across the equator to California. California once had so rich a natural harvest that the Native American tribes had no need to develop agriculture. Today the valleys are some of the most intensively farmed areas in the world.

Cowboys and Indians: California is the birthplace of blue Ceanothus and Californian poppies now seen in our gardens. Out in the spiky chaparral live less familiar faces – scrub oak, buck bush, toyon and many more. The scrub oak, alias the 'chaparro', gave its

name to chaparreros or 'chaps', worn by cowboys to protect their legs when riding through on horseback. Chaparral, grassland and the open oak forests were the results of thousands of years of controlled burning by Native Americans. Today battles are not between cowboys and Indians, but ecologists, conservationists, developers and farmers. Water for irrigation is so valuable that environmental activists have taken out court orders against farmers to make them leave minimum flows in rivers.

Today California has huge levels of resource consumption and wealth accumulation, which, of course, have social and environmental costs. But the region is also the birthplace of innovative new technology and is home to some of today's most environmentally conscious people, searching for solutions to inherited problems.

W·05 Introduction to crops and cultivation

Through the polytunnel and into the kitchen garden of the world: parts of the area you've just visited have sprouted major industries to grow supermarket salads, citrus fruits, olive oil and wines year-round. Huge areas of tomatoes in California, and peppers under plastic in Greece and Spain, need food, water and protection against pests and diseases. Pressure is mounting to reduce subsidies on water and to move to low-input, energy-efficient, self-sustaining and diversified farming – just the kind of crops that a Greek or Spanish grandparent would have used, the ones that made the ancient Mediterranean healthy, wealthy and wise. Back for a future?

W·06 Fruits of the Mediterranean

Loquats and kiwis (Chinese gooseberries) from China, apricots from Iran, and sweet almonds from the Middle East have moved to the Mediterranean and California to soak up the sun – and the water from the irrigation lines. Mediterranean almonds are now one of the most important nuts in world trade.

W·07 Cork

Our cork trees belong to both natural and cultivated landscapes. When *Quercus suber* is 25–30 years old it will do its first strip – for cork. Around 15 billion wine corks are pulled a year, and trees produce around 4,000 corks per strip. Tree destroyer? Definitely not. Cork oaks, unlike most trees, regenerate their bark.

Cork

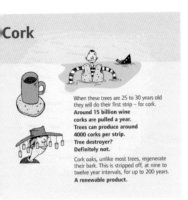

When these trees are 25 to 30 years old they will do their first strip – for cork.
Around 15 billion wine corks are pulled a year.
Trees can produce around 4000 corks per strip.
**Tree destroyer?
Definitely not.**

Cork oaks, unlike most trees, regenerate their bark. This is stripped off, at nine to twelve year intervals, for up to 200 years.
A renewable product.

This can be (skilfully) stripped off at around nine- to twelve-year intervals for up to 200 years without harming the tree at all. The Portuguese Cork Association, APCOR, campaigns to promote cork worldwide.

Cork oak wood pastures can also 'grow' charcoal and meat. Iberian pigs producing high-value ham, *jamon serrano*, feed on the fallen acorns.

These managed wood pastures provide valuable habitats for many plants and animals including the Iberian lynx and 42 species of birds, among them the rare black vulture. So buying real cork conserves the cork trees and the habitat and supports livelihoods. The RSPB works to protect the cork oak woods, and promotes industries that benefit local rural communities.

Heather Jansch, Devon sculptor, used cork and deadwood to sculpt pigs, piglets and the white stork of the open woodlands after visiting the cork oak forests.

W.08 Tobacco

Originally brought to Europe by Columbus, tobacco was heralded as a miracle medicine by many, and as an agent of the devil by a few, centuries before we realized its links to cancer. The tobacco trade was associated with high profit, much of which was linked to government tax revenues, greed, piracy, smuggling and the slave trade.

Tobacco nicotine is more addictive than alcohol, heroin or cocaine. It has been estimated that by 2025 tobacco will kill as many people worldwide as dysentery, pneumonia, malaria and TB combined. 80% of tobacco is grown in the developing world. Tobacco depletes soil fertility, requires many sprays and is a cause of deforestation, but also provides a livelihood for many people.

W.09 Vegetables and herbs

Here you can enjoy displays of Mediterranean produce. Garlic, coriander and basil give dishes a Mediterranean flavour, although basil originally comes from India and the Middle East.

W.10 Citrus

The citrus family is fond of breeding. Clementines are a cross between mandarins and bitter Seville oranges, and tangelos the offspring of tangerines and grapefruits. Citrus are also used in perfumes, cleaning products, anti-bacterial agents and as CFC substitutes. In Florida, citrus plantations need around 127 cm of water a year, double the annual rainfall. This means irrigation. Bore holes are used as sources of river water diminish; some areas have used solar-powered pumps to extract the water at over 1,350 litres per minute. Environmentally friendly? If watering isn't carefully controlled, salts can build up in the soil, reducing yields. Farmers use computers to monitor systems and keep irrigation and fertilization to a minimum, conserving resources and reducing soil salinization.

W.11 Grape vines

Tim Shaw, Cornwall-based sculptor, created this wild Bacchanal – dancing maenads that mirror the twisting shapes of the vines surround their god, Dionysus, here shown as a bull. The vine has sometimes been linked to immortality. Why? Because its short-lived fruits are given long life as wine, because the dead-looking vines burst into life each spring, because we have kept the vine alive through myth and stories for centuries, or maybe because of the effect wine has on us. Certainly, in classical times, mortals were taken to new heights when they drank wine as an offering to the gods.

Dionysus, a potent nature god, was associated with wine, fertility, festivities, intoxication, illusion and destruction. Also known as Bacchus, he started out with good intentions as the god of vegetation. Things changed when he went from growing the vine to drinking its fermented juices ... party time!

The land, like Dionysus, has changed. Here he stands, straddled between the ancient cultivated landscapes of the Mediterranean and the irrigated lands of today's intensive agriculture.

In some Australian vineyards they practise 'partial root drying' where roots on one side of the vine are kept dry while roots on the other side are irrigated. The process is then reversed. The result: fewer side shoots, less pruning and more grapes.

W·12 Peppers

Capsicums, peppers and chillies, which come from the New World, contain a hot compound, capsaicin, in the fruit's inner wall. Measured in Scoville units, mild chillies come in at around 600 units, but the really hot ones get up to 350,000 units. As well as spicing up our food, chilli extracts have been added to sprays, paints and rubber coatings to ward off insects, elephants, aquatic molluscs and rats. Capsaicin is also used in pharmaceutical medicine as an analgesic (pain relief) for neuralgia.

W·13 Grains

Grains feed the world; they are small but carbohydrate-rich and, like pulses, can be dried, stored and easily transported. The major grains are wheat (p.6), rice (pp.6, 30), maize (p.6) and sorghum. Less familiar tef and millet are vital to the diets of local people.

Sorghum bicolor: this African drought-resistant crop is the fifth most important cereal in the world. White-seeded forms make bread while astringent red-seeded types make beer.

Tef: possibly the smallest-seeded grain crop in the world, tef provides a quarter of Ethiopia's cereal and is used to make *injera*, a fermented pancake-like bread.

Pearl millet: an important African cereal adapted to hot, dry conditions, which grows on soils too poor for sorghum and maize.

Finger millet: grows in more favourable conditions. Its tasty seeds, rich in methionine (lacking in the diets of millions), can be stored for years.

W·14 Pulses

For some a tasty treat, for others a means of survival: pulses, edible protein-rich seeds of the legume (pea and bean) family, can be dried, stored and easily transported.

We need nitrogen to make protein. The air is 78% nitrogen, but animals and most plants can't absorb it like this. Peas and beans can by means of bacteria called rhizobia, which live in their roots. In exchange for a home and sugar from the plant, rhizobia turn the nitrogen from the air into plant food.

Certain legumes contain 'anti-nutritional compounds'. Some of these cause indigestion, others can kill. In times of crop failure or drought in Ethiopia, India and Pakistan people often eat grass peas, the only food around. These peas are harmless in small quantities, but over a long period can cause permanent paralysis. The alternative is starvation. Future Harvest's International Centre for Agricultural Research in Dry Areas has now bred a grass pea with low toxin levels that can be eaten without fear of paralysis.

W·15 Sunflowers

Sunflowers are possibly the only major domesticated food crop to have originated in North America, where they provided food, fuel and pigment. We associate sunflowers with the Mediterranean, particularly the South of France, where Van Gogh created sunflower paintings in the 1880s. Today they are grown on the largest scale in Argentina, the Ukraine and Russia, where their oil was first commercially exploited. As well as cooking oil, the seeds provide protein-rich food, margarine, oil for racing-car engines and paint manufacture, as well as having potential for fuel, medicines, cosmetics and plastics. The remaining seed meal feeds livestock, and the husks and stems can be used to fuel the oil extraction.

California is an important centre for sunflower seed production. Look out for the multi-headed male plants in the Californian fields – fathers of tomorrow's sunflowers.

W·16 Tomatoes

The Conquistadors brought tomatoes from South America to Spain in the 16th century. At first the fruits were thought to be poisonous, but the Italians started to eat them after an intrepid herbalist pronounced them edible in 1544. Their reputation as an aphrodisiac comes from a French mistranslation of *pommi de mori* (apple of the Moors) as *pomme d'amour* (apple of love). Nowadays this ubiquitous culinary ingredient is more useful as a source of vitamins A, C and E.

W·17 Olives

Once olive oil provided light for lamps, and the golden essence to anoint the brave, wise and rich and embalm the dead. Today it is used in the kitchen, and is thought to reduce cholesterol levels and deter heart disease. Production is booming, with Spain taking the lead, and the squeeze is on to reduce chemical inputs to keep the land in good heart as well as its people.

Our older olive trees were brought in from Sicily, having reached the end of their productive life in the grove. The younger ones, planted in threes, typify the way table olives are grown in Spain.

Debbie Prosser, local artist, made the olive oil vats from Cornish raw materials. Take a peep inside!

W.18 Plants as perfume

When we give perfume to someone we give them liquid memory – Diane Ackerman

The scent of violets, a whiff of mint – how do they make you feel? Scent goes straight to the seat of emotion and memory in the ancestral core of your brain. Plants use scent to attract pollinators and repel predators. Perfumiers make scents from plant extracts just as musicians use notes to compose melodies. Why do we use perfume? To signal, seduce or warn, like plants, or for sweet memory and comfort? Cleopatra, queen of perfume, power and seduction, wore kyphi (containing rose, crocus and violet) on her hands and aegyptium (almond oil, honey, cinnamon, orange and henna) on her feet. She also scented the purple sails of her barge.

Our five senses, smell, touch, sight, taste and hearing, give us information about our environment. Chemoreception, the detection of chemical signals in the environment, was probably the first sense to appear in primitive organisms, and developed into smell and taste. Humans can now detect over 10,000 different odours. Aromas and perfumed gardens are sometimes used in reminiscence therapy to help people with Alzheimer's. Geranium oil is important and valuable in perfumery.

Quest International, one of the world's leading fragrance companies, combine science with art to create many of the world's best-known scents. Quest have also added scent to the sights in the Biomes.

Following on

Temporary displays, trails and new exhibits

The Warm Temperate Biome is far more seasonal than the Humid Tropics Biome, and every April and October you will see renewed crop displays in the Crops and Cultivation section. This is something which differentiates us from other botanic gardens! In 2004 expect to see sunflowers in here (rather than outside), also maize and pulses that follow on from spring's tulips. In autumn, look out for a new display of dried flower crops and artefacts.

A Little History

They say that success has many fathers, and Eden is living proof of that. The story starts long before our opening on March 17 2001.

The original idea came from Tim Smit's experience in restoring the Lost Gardens of Heligan down the road. No horticulturist himself, he became fascinated with plants and our relationship with them. Could a greenhouse and a site be found in Cornwall big enough to tell the story?

What persuaded two heavyweight horticultural prize-fighters, gently settling into well-earned pre-retirement mode – Philip McMillan Browse (former Director of RHS Wisley Horticultural Director of the Lost Gardens of Heligan) and Peter Thoday (of *Victorian Kitchen Garden* fame and President of the Institute of Horticulture) – to get back to fighting weight and bring together as fine a horticultural team as could be found anywhere? What made our local council, the Borough of Restormel, put up the first £25,000 and give this story a beginning?

And what of the architects? The first architect involved was Cornishman Jonathan Ball, who worked with Tim as co-founder of the Project and then passed the baton to Nicholas – now Sir Nicholas – Grimshaw. The Grimshaw practice it was who stepped up to that now-famous challenge: 'The good news is that we're giving you the chance to build the eighth wonder of the world. The bad news is that we can't pay you.' Eventually, of course, we did pay, and Grimshaws not only designed the fabulous buildings already on site but are currently working on Phase 4 (see Big Build 2 on p.18).

How is it possible that two hard-nosed construction companies, Sir Robert and Alfred McAlpine, worked for 18 months without payment or contract and then, for good measure, agreed to loan Eden a significant sum only to be repaid if the Project was successful?

For three years our team battled away at making the dream become reality, and today £100 million has been invested in this pit. The Millennium Commission weighed in with £37.5m of lottery funding to single Eden out as the 'landmark' project of the far South-West, and their subsequent contributions will bring the total to just over £50m. We hope we've delivered for them and for anyone who ever bought a lottery ticket and casually wondered where the spare money would go once they'd pocketed their jackpot. Other major sources of funding included the European Commission and South West Regional Development Agency; a full list can be found on p.60. Come to that, what are the chances of a public project in Cornwall being lent £12m by the NatWest Bank, especially bearing in mind that no other Millennium Project has had bank support? An imposing phalanx of bankers, accountants and lawyers were persuaded to support an incredibly complicated finance plan that was to bring in another £6m.

Together, our funders took a massive risk on a massive dream, and it became their dream just as much as ours. Vast numbers of people who might have found it much easier to say no said yes – and the rest is history.

The **Eden Trust**

The Eden Project is owned by the Eden Trust, a company limited by guarantee and a UK Registered Charity – number 1093070.

The Eden Project is the physical home for the Eden Trust and its objectives in the areas of education, horticulture, the environment, conservation and sustainability. We aim to

* break down the barriers to communication, sharing information and ideas with the widest possible audience;

* explore the potential for working with the grain of nature, bringing together science, art, technology and commerce to create a constituency for change – then help to put it into action.

All profits go to further the charitable work of the Trust, and we continue to fundraise to supplement this income.

A **foundation** for **the future**

The original Garden of Eden is a symbol of paradise, but also of mankind's ejection from it. Historically, conservation policies have assumed that the best-quality environments are those untouched by people, and that environmental care meant keeping people out of them. The real situation is more complex.

Humans have caused problems in the world, but there are also places where we have lived in harmony with nature without complete destruction, and sometimes with a beneficial effect. Communities have already begun to make steps to be effective stewards of the world. The Eden Project is here to showcase those steps, and to tell how each one of us is, already, a global citizen.

We're also here to show that environmental awareness is about quality of life at all levels. The 'environment' is shorthand for issues that impact on us in a thousand ways every day, from the food we eat and the clothes we wear to the weather we enjoy or suffer. Understanding our world better, and the part we play in it, is also about having fun, not about living grey, hair-shirt lives. Our gardens and displays are used as a lens to focus in on the amazing worlds that each one represents; how the politics of the world, for example, lie within a cup of sweet tea. Our plants were chosen because they could tell a story, and every story is there; gruesome, awesome, funny, inspirational, telling of amazing science, giving us hope for food, security, clean technologies and improved health.

We have created wonders to enjoy here in Cornwall, but of course they are just echoes of the real world's diversity, and of the lives that people live in these places. We look forward to showing you where we have got to next time you visit.

The Eden Foundation underpins everything we do. We will not waste time and energy doing what others already do well, so we are working with a wide network of partners to complement their work. Together we want to tackle the big, difficult questions. We want to identify the barriers to a better understanding and a better world, and start to break them down. The Foundation will be the crucible where we explore approaches, make new alliances and dare to experiment with ideas that may have wider value. It will not be easy, it will not all work – but the bits that do will be fabulous. This is our Project, this is your Project – always in evolution. We do hope you will enjoy joining in.

The **Biomes**

These iconic, super-efficient Biomes are the biggest conservatories in the world, as confirmed by the 2004 *Guinness Book of Records*. The Humid Tropics Biome, which could hold the Tower of London, is 11 double-decker buses high and 24 long, with no internal supports. Building these 'lean-to greenhouses' on an uneven surface that changed shape was tricky. Bubbles were used because they can settle perfectly on to any shaped surface. The bubbles were made of hexagons, copying insects' eyes and honeycombs – nature's common sense, producing maximum effect with minimum resources. The Biomes' steelwork weighs only slightly more than the air they contain. They are more likely to blow away than blow down, so are anchored into the foundations with steel ground anchors: 12-metre tent pegs!

The final design comprised a two-layer steel curved space frame, the hex-tri-hex, with an outer layer of hexagons (the largest 11 m across), plus the occasional pentagon, and an inner layer of hexagons and triangles (resembling huge stars) all bolted together like a giant Meccano kit. Each component was individually numbered, fitting into its own spot in the structure and nowhere else.

The transparent foil 'windows', made of 3 layers of ETFE (ethylenetetrafluoroethylene-copolymer), form inflated 2-metre-deep pillows. ETFE has a lifespan of over 25 years, transmits UV light, is non-stick, self-cleaning and weighs less than 1% of the equivalent area of glass. It's also tough: a hexagon can take the weight of a rugby team. The pillows were installed by 22 professional abseilers – the sky monkeys.

The **earth**

We made 85,000 tonnes of topsoil because we didn't want to deprive anyone else of theirs and wanted to get the recipes right. Soil is made of minerals of different sizes (sands and clays) and organic matter, mixed together in the right proportions. Soil manufacture has been tried before, but nowhere with so much at stake. The team worked with partners at Reading University and within a tight budget and time-frame using wastes and recycled materials where possible, and avoiding peat. Local mine wastes provided the minerals; the china clay company IMERYS had a bit of sand spare, and WBB Devon Clays Ltd had some reject clays.

The organic matter needed to be tough and long-lived, especially in the Biomes, where composted bark from the forestry industry was used. Plants in the Humid Tropics grow rapidly and needed a high-performance soil capable of holding lots of water and nutrients. In the Warm Temperate we put more sand in the soil so that it holds less water and nutrients to show how plants function in their natural, dry, harsh environment. Bark and clay made up the rest of the mix. For the South African Fynbos plants fertile soil is toxic. This mix therefore had to be almost nutrient-free, and consists simply of composted bark and sand. Outdoors, where the climate is less demanding, we went for composted domestic green wastes. The ingredients were mixed together with a JCB in a nearby clay pit, like making a giant cake. Worms from Wiggly Wigglers helped to dig and fertilize the new earth.

Making our soil demonstrates an important part of our work; that environmental regeneration is possible. It is also something that now has an application in the wider environment and that we hope will assist regeneration projects far beyond Bodelva.

he **climate**

Eden's artificial Biome climates are constantly monitored and controlled automatically.

Mist and rain: In the Humid Tropics automated misters moisten the air (90% relative humidity at night, and down to 60% during visiting hours) and ground-level pipes irrigate the soil so you don't have to put up with the rainforest's 1,500 mm (60 inches) of rain a year! Humidity is also helped by our crashing waterfall, whose water is all recycled, of course, brought to us with the help of Pennon Group Ltd. and their subsidiary, South West Water Ltd. In the Warm Temperate we keep moisture levels down; vents are often open, even during relatively cool periods, to reduce humidity close to the leaves (which can otherwise cause fungal problems).

Hot and cold: The main heating source for both Biomes is the sun. The back wall acts as a heat bank, releasing warmth at night. The two layers of air in the triple-glazed windows give maximum insulation. Extra heating is provided through the air-handling units, the big grey boxes outside the Biomes. The Humid Tropics Biome ranges from 18°C to 35°C; and the Warm Temperate reaches 25°C in summer with a winter minimum of about 9°C.

Ventilation: The vents may seem small for a building this large. They work because the height of the Biomes generates a 'chimney effect' that draws air through the system. On very hot days the air-handling units help circulate the air within the Biomes.

The **animals**

Pollinators: We only need to pollinate the flowers if we want them to produce seeds or fruits. Many of our plants are wind-pollinated and many of the insect-pollinated ones accept the services of any passing insect. In special cases, however, we use a member of staff with a paintbrush!

Pest control: We run a rigorous healthcare programme to control pests. Isolation houses at Eden's nursery catch problems before they reach the pit. On site we use an integrated pest-management system. In the Biomes we have introduced DEFRA-licensed biological controls including birds, frogs and lizards as well as beneficial insects. The animals came from Newquay Zoo and have been bred in captivity. To make sure they stay healthy and safe we would be grateful if you do not feed or handle them.

'Soft' chemicals (soaps and oils) are used as required. Strings dangling from the taller trees are small pulley systems that insert beneficial insects into the canopy. In some areas lightboxes emit UV light at night to catch moths and mosquitoes.

Sustain-ability...

... is the ability to minimize the impact of our actions and maintain things at a steady level without exhaustion. When making any decisions at Eden about tackling waste, water, energy and the day-to-day impact of the whole Project, we take environmental and social as well as economic considerations into account.

We're doing this in an open way that makes our buildings, our practices and our organization part of the education to help others, both organizations and individuals, think about their impacts too. For many of our challenges there are no easy answers, and every issue requires vigorous debate. That isn't an excuse, but it is an important process that will enable us to make the right long-term decisions.

We aim for efficiency wherever possible – using less is one of the greatest steps towards sustainability. Our Biomes are among the most efficient structures ever seen, using minimal materials that are long-lived and easily recyclable and use only a third of the engineers' estimates for energy to heat them. We source our supplies locally wherever possible, reducing transport energy and supporting local economies; our soils reuse waste; our water is predominantly recycled groundwater and harvested rainwater. We buy our Green Tariff electricity from Cornish wind farms, run many of our vehicles on LPG and have a green travel plan for staff and visitors, an important element of the government's integrated transport policy aimed at promoting greener, cleaner travel choices.

We have now developed a programme called **Waste Neutral**, where we:

1: reduce waste.

2: reuse items wherever possible.

3: source all remaining items, wherever possible, from materials that can and will be recycled.

4: adopt a policy of purchasing items that are made from recycled materials, either for use on site or for sale in the shop.

In simple terms, when we buy in a greater weight of products made from recycled materials than the weight of materials we send off to be recycled we will have reached Waste Neutral. This concept can be applied to any organization, community or even individual household.

To become Waste Neutral we need your help. Please use the recycling bins around the site. A major obstacle to recycling is contamination of waste streams – if you throw a sandwich in the plastics or glass bin we have the messy job of getting it out. If you can't find an appropriate bin then please use the general litter bin. We are in the process of building a permanent recycling compound in Pineapple car park where you can discover how we recycle paper, glass, metal, cardboard, wood and wood waste, and watch our organic waste turn into beautiful, useful compost in our biodigestion unit.

Help us to reduce, reuse, repair and recycle. Get involved.

The Eden teams

We started with 5 people, now we are over 500. It's a bit like an ant colony, with a number of teams all of whom work with each other so the whole is greater than the sum of the parts. The teams report to the Board, who report to the Trustees, who ensure we meet our charitable aims.

The **Foundation Team** work with partners and supporters worldwide to act as a window on the world. The Foundation explores ideas and develops projects, education programmes, creative and interpretation programmes that marry art, science and technology whilst keeping a close eye on the integrity of the project.

The **Marketing and Media Team** take us out there, help to decide how we look and how we communicate, publish books, run Friends and research the lot to make sure we're going in the right direction.

The **Development Team** look after our people, our projects and our commercial and technological development, building an enterprise that transforms ideas and ways of doing business.

The **Destination Team** bring it all to you: the landscape and plants, the exhibits, the events, the retail, the catering and the housekeeping: nothing is franchised out, it all links together: what we grow, cook and sell and the way we act is as much a part of the story as what the world out there is up to.

The **Resource Team** make sure we run a lean machine, spend wisely and keep our feet firmly on the ground. It's no good talking regeneration and positive futures unless we demonstrate good practice here and now.

The **Visiting Team**: that's you. You are part of Eden and part of the story, and your comments and views are always appreciated.

rt and science

In true gardening style we do a lot of planning, digging, sowing, growing and cross-pollination of ideas. Take art and science for example: art at Eden is not a substitute for facts and information. The artworks are signposts to new attitudes and ways of thinking. They are not always comfortable, they are not always beautiful, but they should always be surprising and thought-provoking. We are working with a wide range of artists, locally, nationally and internationally, who, along with our in-house team of multi-skilled Designer Makers, breathe life and colour into the Project.

Creating Eden's exhibits and growing our plants requires a huge scientific input from a range of disciplines. Scientists and the Green Team work together to ensure that we display top-class living plant material from which the stories can be told. The team also helps to research and authenticate Eden's stories and messages. Scientists work hand-in-hand with communicators and artists at Eden and come up with some fantastic results, such as the Flybot (see p.37).

Where has the money come from?

Maintaining a strong and diverse financial base is crucial to preserving the Eden Trust's independence and credibility. None of our work, none of our successes, would be possible without the generosity of our donors and supporters. Thank you.

Arts Council, South West; Arts Council, New Audiences; The Misses Barrie Charitable Trust; The Body Shop Foundation; John Coates Charitable Trust; CISCO Foundation; Clipper Teas; The Ernest Cook Trust; COPUS: Connecting People to Science; Cornwall College; Cornwall Arts Marketing; Cornwall/Aylesford Paper; Cornwall County Council; Cornish Horticultural Enterprises; Cornwall Catering; Creative Partnerships; Creative Kernow; The Darwin Initiative; Day Chocolate Company; The Department for International Development; The Department of Education & Skills; DEFRA; Devon and Cornwall Training and Educational Development Council; Dyneon Gmbh – A 3M Company; English Nature; English Partnerships; The Environment Agency; Ernst and Young; Estée Lauder; European Agriculture Guidance and Guarantee Fund; European Regional Development Fund; European Social Fund; Founder Friends and Friends of Eden; The Garfield Weston Foundation; The Headley Trust; Kellogg's; The Kleinwort Benson Charitable Trust; Landfill Tax Credit Scheme; Lloyds TSB Foundation; Millennium Commission; Mobile Infrastructure; The Nationwide Foundation; NESTA – National Endowment for Science Technology and the Arts; Pennon Group plc; Restormel Borough Council; Rio Tinto; Rural Development Commission; Simon Robertson; Single Regeneration Budget; South West of England Regional Development Agency; South West Water Ltd; Syngenta Foundation; The University of Reading; Viridor Credits; Visiting Arts; Waitrose Ltd; The Wellcome Trust – PULSE Engaging Science

Working together

We are privileged to have some of the most forward-thinking and effective organizations working with us to bring our 'Living Theatre' to life. We are delighted to work with those who are determined to foster improvements in lives, livelihoods and the environment. In the coming years you will see our supporters taking to the Eden stage, sharing their work, their ambitions, their concerns and, we hope, something of the excitement of making a difference. They come from a wide range of disciplines and have a wide range of views. Eden is a place of many voices – including, of course, yours.

Some of the key people who have supported the development of our exhibits and site are listed below. See others during the year, especially as part of our Eden Live programme.

Biscuit, Cake, Chocolate and Confectionery Association (www.bccca.org.uk); **Body Shop Foundation** (www.the-body-shop.com); **BT** (www.bt.com); **CABI Commodities** (www.CABI-

Commodities.org); **Cafédirect** (www.cafedirect.co.uk); **CISCO** (www.cisco.com); **Clipper** (www.clipper-teas.com); **Combined Universities in Cornwall, School of Geography, Earth Sciences and Archaeology (Camborne School of Mines)** (www.cuc.ac.uk); **Cornwall College** (www.cornwall.ac.uk); **Cornwall County Council** (www.cornwall.gov.uk); **Creative Partnerships** (www. creative-partnerships.com); **Darwin Initiative** (www.darwin.gov.uk/); **Duchy College** (www.cornwall.ac.uk/duchy); **English Nature** (www.englishnature.org.uk); **Environment Agency** (www.environment-agency.gov.uk); **Environmental Protection Agency (EPA), Guyana. Iwokrama Field Centre and the people of Fairview** (www.sdnp. org.gy/epa); **Ethnomedica** (www.rbgkew.org.uk/ethnomedica); **Fairtrade Foundation (FTF)** (www.fairtrade.org.uk); **Falmouth College of Arts** (www.falmouth.ac.uk); **Fauna and Flora International** (FFI; www.fauna-flora.org); **Forest Stewardship Council (FSC)** (www.fsc-uk.org); **Future Harvest** (www.futureharvest.org); **Global Canopy Programme** (www.globalcanopy.org); **Green and Black's** (www.greenandblacks.com); **The Guardian** (www.learn.co.uk); **Housings and Hazards, Exeter** (www.HazardResistantHousing.com); **IMERYS** (www.imerys.com); **Infapro, Cameroon, Danum Valley Project** (www.ssl.sabah. gov.my/project/danum.htm); **International Institute for Environment and Development** (www.iied.org); **Kneehigh Theatre Company** (www.kneehigh.co.uk); **Landlife** (www.land life.co.uk); **Lost Gardens of Heligan** (www.lostgardensofheligan.co.uk); **MAKHAD Trust** (www.makhad.org); **National Non Food Crop Centre** (www.nnfcc.co.uk); **Natural Resources Institute (NRI)** (www.nri.org); **Natural Resources International Ltd** (www.nrinternational.co.uk); **NESTA: the National Endowment for Science, Technology and the Arts** (www.nesta.org.uk); **October Gallery in London** (www.october.gallery. ukgateway.net); **Pennon Group plc** (www.pennon-group.co.uk); **Plantlife** (www.plantlife. org.uk); **Portuguese Cork Association (APCOR)** (www.apcor.pt); **Quest International** (www.questintl.com); **Rainforest Alliance** (www.ngowatch.org/rainforestalliance.htm); **Rainforest Concern** (www.rainforestconcern.org); **Raleigh International** (www.raleigh. org.uk); **Rio Tinto** (www.riotinto.com); **Royal Botanic Gardens, Kew** (www.rbgkew. org.uk); **Royal Society** (www.royalsoc.ac.uk); **Royal Society for the Protection of Birds** (www.rspb.org.uk); **Scottish Crop Research Institute (SCRI)** (www.external. scri.sari.ac.uk); **Sensory Trust** (www.sensorytrust.org.uk); **Seychelles Ministry of the Environment** (www.virtualseychelles.sc/gover/me.htm); **South West Arts** (www.swa.co.uk); **South West Water Ltd** (www.swwater.co.uk); **St Helena National Trust** (www.sthnattrust.org); **Technoserve NGO in Ghana** (www.technoserve.org); **The Guardian** (www.learn.co.uk); **Timber Research and Development Association (TRADA)** (www.trada.co.uk); **Tomato Growers Association** (www.britishtomatoes.co.uk); **University of Bristol** (www.bris.ac.uk); **University of Exeter, Camborne School of Mines** (www.exeter.ac.uk/CSM); **University of Exeter, School of Education** (www.exeter.ac.uk/ education); **University of Kent at Canterbury** (www.kent.ac.uk); **University of Plymouth** (www.plymouth.ac.uk); **University of Reading** (www.reading.ac.uk); **University of Sheffield** (www.shef.ac.uk/landscape); **University of the West of England** (www.uwe.ac. uk/fas); **Viridor** (www.viridor-waste.co.uk); **Visiting Arts** (www.britishcouncil.org/ visitingarts); **Waitrose** (www.waitrose.com); **Wateraid** (www.wateraid.org.uk)

What can you do?

You've now discovered some of the diversity that keeps you alive: plants for foods, fuels, medicines, tools; plants from the wild places that clean our air, moderate our climate and refresh our minds; the people who grow our crops and conserve our land. How do we sustain the life that sustains us? Your choices matter. How you live, what you buy, what you decide – you do make a difference. Join us ...

Friends

Eden Friends now has a membership of 10,000 people from all over the world – active partners in our future. They enjoy unlimited free entry, a quarterly magazine, a programme of special events, talks and workshops and 10% off all retail purchases.

Membership levels are: **Individual** £35 p.a. **Joint** (2 people at same address) £60 p.a. **Family** (2 adults and up to 3 children in full-time education) £70 p.a. **Lifelong** £1,000

Or save £100 with a **ten-year membership**: Individual (10 years) £250 Joint (10 years) £500

For more details please phone 01726 811932, visit the Friends' desk in the Ticketing Hall or email **friendsdesk@edenproject.com**. If you join during your visit your admission money will be refunded.

Pick up a Passport

For just £5 administration fee on top of the price of entry your Eden Passport will entitle you to free admission for a year from the date of your visit. To get your Passport keep your till receipt and fill in an application form, available from the information desk.

ant to help further?

There are other ways to help us. For information about making a donation, whether once only or by regular direct debit, about working as an Eden Volunteer in horticulture, visitor support, administration, fundraising or education, or about including a legacy to Eden in your will, please contact the Fundraising Team (**LCompton-McDonald@edenproject.com** or at the usual Eden address).

ant to know more?

Leaflets are available around the site, and more information is available at our website, www.edenproject.com. We also publish a range of Eden Project books for adults and children of all ages, which are available at the shop in the Visitor Centre, and by mail order from **www.edenstore.co.uk.**

ank you ...

We would like to say a big thank you to all of you who have supported us over the years. The list grows so rapidly that it is sadly impossible to mention everyone here, but you are all on the web. You know who you are, so if you are not on the list, and should be, please let us know.

st but not least

Thanks again to the Eden Team and the entire construction, design and professional teams who have given beyond the call of duty and whose talents are imprinted right across the site.

First published 2001 by Eden Project Books
a division of Transworld Publishers

Fourth revised edition 2004

Text and design © the Eden Project/Transworld Publishers
2001, 2002, 2003, 2004

Text by Dr Jo Readman with assistance from the Eden team

Transworld Publishers, 61-63 Uxbridge Road, London W5 5SA
a division of the Random House Group Ltd

www.booksattransworld.co.uk

ISBN 1 903919 30 4

Editor: Mike Petty   Design: Charlie Webster
Site maps prepared by Gendall Design

All photographs copyright

Photo credits: Bob Berry, Peter Blackburn, Philip McMillan Browse, Simon Burt, Tony Carney (Apex), Duncan Carr, Alan Clarke, English Nature, Fairtrade Foundation (Apex), Charles Francis, Robin Fuller, Future Harvest Centres, Nick Gregory (Apex), Nicholas Grimshaw and Partners, Richard Kalina, Tony Kendle, Glen Leishman, Robin Lock, Sue Minter, Andrew Ormerod, Mike Petty, Plantlife, Glenys Pritchard, Jo Readman, Laura Richardson, Monroe Sheppard, Paul Spooner, Steve Tanner, Charlie Webster. Front cover: Winston Woodward.

Apologies for any omissions – we update the guidebook regularly, so if we have not credited you above please let us know and we will make sure we include you next time.

Printed in Great Britain by Butler & Tanner Ltd, Frome,
on paper from a managed, sustainable forest using
a process that is totally chlorine free

The Eden Project is owned by the Eden Trust, registered charity no. 1093070 and all monies raised go to further the charitable objectives.

Eden Project, Bodelva, St Austell, Cornwall, PL24 2SG
T: +44 (0)1726 811911   F: +44 (0)1726 811912

www.edenproject.com